Copyright © 2020 J. KORSTANJE

All rights reserved.

ISBN: 9798642734674

GUIDE TO SUCCESSFUL RESEARCH

This book is dedicated to all things misunderstood.
It would be boring without them.

GUIDE TO SUCCESSFUL RESEARCH

CONTENTS

GUIDE TO SUCCESSFUL RESEARCH

THE 8 RULES OF RESEARCH

The reason you are reading this guide is probably that you want to do a good research. So how to start any better than by understanding what good research actually is!

Let us depart from the following:

Research is an objective approach to finding a general truth to a question.

A great analogy is making a 1-million pieces puzzle. Solving a complete puzzle would be very long and complicated. But finding the position of one puzzle piece would be a relatively acceptable task.

We can see each individual research as a solution for a piece of a puzzle, while the whole puzzle is being solved by the complete research field working on your domain.

So each research is a small contribution to a larger picture.

Making a research as a puzzle piece seems easy at first, but it is a challenge to make it fit perfectly: firstly, it should be not too large nor too small. But when you place a puzzle piece, you should also be a hundred percent convincing about its position, to not confuse all other puzzlers.

By following the 8 Rules of Research of this guide, you will make sure that your research is a success!

THE 8 RULES OF RESEARCH

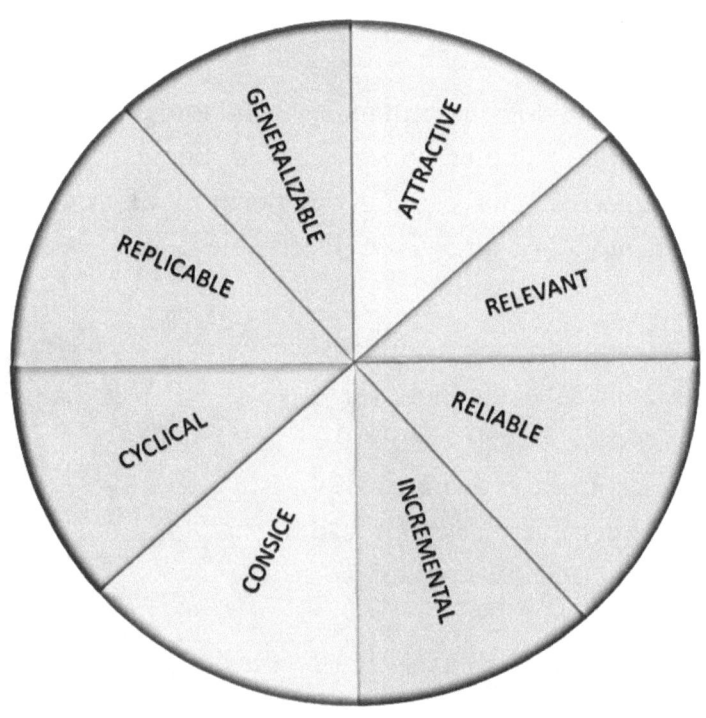

RULE 1 – BE ATTRACTIVE

What is attractive research?

Attractive research is about making sure that your research will attract readers.

Imagine you spend months working on a paper, and when you finally finish it you find out that there's actually no one to read it. That would be a big waste of time!

How to make your research attractive?

Tip 1 - The topic of your research

No-one will tell you this, but the topic of your research is far more impactful than you'd think. There is a huge fashion aspect to scientific work.

It is a bit anti-scientific to take fashion into account when choosing a research topic, but everyone does it. To justify this, one could even argue that a hot topic only becomes a hot topic if it is in desperate need of scientific contributions.

Fashion is also a way to make sure your research will be useful, and, besides that, it is much better for your career.

Tip 2 – Make it likely to have a positive outcome

Most research poses a research question with a hypothesis and then tries to find proof of this hypothesis.

The conclusion can be:

- Positive: you have found proof for your hypothesis
- Negative: you have not found proof for your hypothesis

The difficulty is here: in the positive case, you prove that the hypothesis is correct, but in the negative case *you do not prove* that the hypothesis is wrong: you just have no definite conclusion.

Both negative and positive conclusions add knowledge to the scientific domain you work on, but unfortunately nobody really cares about negative conclusions.

Negative conclusions are not likely to attract much readers and therefore will not make you successful. Take this into account in the choice of your research.

There is no harm in choosing a smaller research topic on which you are more likely to find a positive conclusion. It is more valuable to add a small bit of knowledge of which we are sure, rather than trying to find a larger bit of knowledge but failing in validating it.

Good and Bad Example of Attractiveness

1 - Climate Change and the Great Barrier Reef

A great example of attractive research at this moment is a research on the effect of global warming on the Great Barrier Reef.

In addition to combining two topics that interest a lot of people, the study has found proof for their hypothesis that the coral reefs are changing due to global warming.

It being impossible to launch a set of experiments on the great barrier reef, it is very hard to prove solidly that the one and only cause of the change in coral reefs is global warming. But thanks to the great choice of research topic, the authors were able to get a great publication to a lot of readers.

2 – A Cure for a disappeared epidemic

The reason for studying attractive topics is obvious. If the topic is attractive, you can make useful contributions with direct implementations.

However, there is a counterpart to this. Sometimes, irrelevant topics are researched, then left aside for years and years, only for a rare situation to make them very hot topics all at once.

An example of this is a study in medicine. The study started when the SARS lung diseases started spreading around the globe quickly. The researchers found a medicine against it, but only when the epidemic was already over.

But when, 10 years later, another epidemic spread around the world (Coronavirus), the study finally reached its top priority as the Coronavirus was very similar to the SARS virus and the medicine could be potentially re-used.

RULE 2 – BE RELEVANT

What is relevant research?

Relevant research is research that is useable. Depending on your research, this can come from a theoretical contribution or from practical applicability.

An example of relevance in research is if two concurrent researches find different solutions for a same problem. If both solutions are true, but one is more practically useable, it will be more relevant and more successful.

How to make your research relevant?

Tip 1 - The goal of your research

To make your research relevant, make sure that the goal of your research is well stated before starting. Before doing any research work, state a goal in which you decide which type of added value you are going to deliver.

For many, stating the goal seems a useless task. But having a goal allows you to test at every step of your research whether you are not losing track along the way, which happens quite regularly.

So during your research, for every addition you make, you should ask yourself whether this addition answers to the research goal.

7

Tip 2 – Personify your research goal

A great way to help with your research goal is to imagine a person that would be the most important public for your research.

Having this person in mind allows you to reason practically about how you expect this person to benefit from your research and whether it would really have an added value for him.

Tip 3 - Preview conclusion of your research

A third tip to apply for being relevant is to pre-define the scope of possible conclusions can have. This is something you can best do at the stage where you define the research questions.

As soon as you have defined these questions, you should be able to identify all possible answer, possibly with a little bit of speculation.

Then combine this with tip 2 and check whether all of the potential outcomes would be relevant for your identified public. If not, try to tweak the questions in such a way that all possible outcomes would be relevant. This will increase your chances of a successful research.

Good and Bad Examples of Relevance

1 – A Relevant Example

As example of a highly relevant research, let us look at a 2016 paper in the domain of Artificial Intelligence. The paper proposed a method of recognizing objects on images.

The domain of Artificial Intelligence is highly competitive as many researchers will test new methods for the same tasks and compare performances.

Proposing a new method for a same task, together with the performance it has obtained, generates highly relevant research. Whether the performances are good or bad, it will inspire other researchers to find direction for other tests and explorations. And if the performances are good, the method can be used directly in practice.

2 – Psychopaths like bitter tastes

A 2015 research has examined the relationship between taste preferences and psychopathic personality traits. The conclusions were positive and became trending on several popular news websites.

Even though it is very interesting to know that psychopath behaviors are statistically related to

9

preferring bitter taste, this information is not very practical: preferring bitter tastes does in no way guarantee that you are a psychopath.

RULE 3 – BE RELIABLE

What is reliable research?

Reliable research is research in which others can trust. The reliability of a research is an important factor in its usability.

Going back to the comparison of Science as a Puzzle, when a research adds a puzzle piece to the larger puzzle, it must be certain that the puzzle piece is going to the right place and the research must succeed to convince other puzzlers about that.

How to make your research reliable?

Tip 1 – Do hypothesis-based research

The most reliable way of doing research is to start your work from a literature research on a topic, in which you identify hypotheses for your research question.

Before starting to collect data, you should have already identified which data would support which hypothesis. Working in this order ensure to obtain results that can be easily checked by peer researchers on all sides.

Tip 2 – Make sure your data is obvious proof

In the scientific approach, your creativity is going to be in the definition of the hypothesis. You should ensure that your hypotheses are well-rooted in the current state of the art of science and that the data you collect is obvious proof or anti-proof of those hypothesis.

If research method is set up badly, you may collect data that can be interpreted in different ways and therefore not solidly prove or un-prove a hypothesis. In this case you can only write a reflection on the difficulty of interpreting your data and the research will probably not be successful.

A good tip is to try to argue the opposite conclusion for the same data. If there is a way to argue the opposite, this means your conclusion is not obvious from the data and you risk an interpretation bias. This will not be a reliable paper.

Tip 3 – Measure the right thing

A last tip to consider for reliable research, is to measure what you want to measure. This seems obvious, but in practice it often goes wrong.

Many researches are based on surveys that ask many people their opinion. In many cases, people are asked things that they cannot answer reliably. For example, when asking whether people would rather buy a product A or a product B, they may well think that they prefer product A, but as soon as they are in the store, they buy product B. The reliable way to do this research would be to not use a survey, but rather to set up an experiment with a store and see which product is bought more often.

Good and Bad Examples of Reliability

1 – A Reliable Example

An example of very reliable research is so-called A/B testing. It is a very popular research method among website designers in which they make two alternative versions of a website and they compare the number of sales in each of the versions of the website.

The reliability if this research comes from the fact that no-one could interpret the conclusions in any different way than it is: one of the versions made more sales and therefore that version of the website is plainly and obviously better.

2 – An Unreliable Example

As an example of unreliable research let us take a student thesis on cat food. This student did a research for a cat food company that wanted to find out how their company's image came across to the public.

The student had not time to waste and proceeded quickly to launch a survey on the internet, asking what people taught of the company with an open question.

After having collected hundreds of respondents, the student went on to analyze the results and found out that people gave such different answers. No problem, thought the student, and he grouped them in three groups and presented this to the company.

The presentation went well until out of the public came the question: "Why did you choose those groups? If you define the groups differently, your research would be the opposite." The lack of a methodic answer had made this research unreliable and it would not be possible to use it.

RULE 4 – BE INCREMENTAL

What is incremental research?

Incremental research is research that adds new knowledge but bases itself on previous papers. Research can sometimes be done to validate conclusions of prior research, but usually new knowledge is added.

It is important to keep the balance between being new and being funded in prior research. Research that is very innovative often lacks foundation in mainstream knowledge and is unlikely to be accepted. On the other side, if the conclusions are too close to prior existing knowledge, your research is unlikely to attract much interest.

How to make your research incremental?

Tip 1 – Do exhaustive literature research

The most important tip for being incremental is doing very extensive literature research. In order to be incremental, you need to know your domain by heart. Investigate everything that has been done before, but also look at research projects that are work in progress. You need to know exactly what has been done before, to be able to find a hole in the existing knowledge that you can fill with your research project.

15

Tip 2 – Collaborate with recognized researchers

If you are doing very innovative research, it is important to have a well-recognized researcher on your side. If your conclusion is not in line with the mainstream knowledge, you should be prepared for a lot of criticism and it will strongly help you if you have succeeded to convince a peer researcher that has important status in your domain.

At the same time, experienced researchers of the domain will be able to direct you to existing resources and help you to make your work more relevant.

Information that a well-connected researcher could give you is for example information on concurrent projects that are not yet published or failed projects that have never led to anything interesting. Since unsuccessful projects are usually not published, this type of relations can help you a lot.

Good and Bad Examples of Incrementality

1 – An Incremental Example

As a good example of incremental research, let us go back to the example that was previously listed as a bad example in the part on relevant research: it showed that psychiatric personality traits are related to preferring bitter tastes. It was said that this information would be little relevant as we cannot class someone as a psychopath based on a preference for bitter taste.

Being relevant and being incremental can sometimes be conflicting. Going back to the puzzle comparison, the most successful research is one that succeed in finding a large puzzle piece very convincingly. But starting such a large research project is risky, because you might as well find nothing, making your research unattractive.

If you make sure to be funded in other research and add a small but reliable increment of knowledge, you take the low-risk route of delivering a decent scientific contribution, which is always better than trying to find something huge and not succeeding.

2 – A Non-Incremental Example

As an example of a non-incremental research let us take a social science research student project that tried to look at changes in political attitude on a specific topic. The student used a quantitative method while the reviewers were all very much grounded in qualitative research.

Even though the research was not done wrongly, it was a total mismatch with prior research methods in the field. Having not based the research enough in previous work, it was very difficult for reviewers to appreciate the work and it therefore received bad feedback.

RULE 5 – BE CONSICE

What is concise research?

Being concise is about two things: on one side, there is concise factual writing and at the same time there are ways to be concise in the content. You must be concise in what you present to a reader and at the same time in how you present it.

How to make your research concise?

Tip 1 – Avoid page filler, but also missing details

Writing styles usually differ per person. Some may have a hand of not giving enough context. Others may be too talkative in their writing.

It is important to understand that writing is research paper is not like writing a novel. Your readers expect to find every necessary detail, without finding any unnecessary details.

Scientific writing being a very logical style of writing, a good tip is to check your text to find out if every sentence has additional value. If not, you may want to reword your sentence and find a more factual representation of that message.

Tip 2 – Identify your message before starting to write

A second tip for concise writing is to do a mapping of the messages that you want to give to the reader per part of your writing.

A great example is to write down all titles for your chapters and subchapters first. Then check whether the logic is clear and when it is, start adding your different findings as bullet points to the chapter titles.

You should have findings in each subtitle, or else you'd better remove the empty titles. The last thing is to convert the bullet points in sentences and add some langue to make it readable.

Tip 3 – New question, new research

The last tip for being concise is to look critically on what you are writing or researching and to be able to judge certain topics out-of-scope. As a research topic often raises many other ideas and hypothesis, it can be a risk to address them too early in your research and make the paper confusing.

The research should be answering only one question. And questions and hypotheses generated by this should be listed only in the final reflection chapter of your paper.

Good and Bad Examples of Concise

1 – A Concise Example

As examples of concise and unconcise research, here are two ways of writing a report of the same experiment in which a new medicine is tested on two groups of people.

Two treatments have been applied to groups of randomly drawn participants. Treatment A has been found to generate significantly higher survival.

---

2 – An Unconcise Example

As an unconcise example, look at the following text that gives some details that do not add value to the reader:

As discussed before, we have two treatments in this research: medicine A and medicine B. Since we wanted equal groups to test this on, participants were divided in two groups: one group for treatment A and one group for treatment B. What we found was a higher survival in group A. In order to validate it, we also did a statistical test that confirmed this finding statistically.

RULE 6 – BE CYCLICAL

What is cyclical research?

The scientific approach to research is to start your work from a literature research on a topic, in which you identify a gap of knowledge that you try to solve in your work.

Cyclical research is very related to incremental research. There where incremental research is research that builds further on previous work, cyclical research is research that paves the way for future research.

How to make your research cyclical?

Tip 1 – List unanswered questions of your research

A first way to be cyclical in your research is to clearly list all unanswered questions in the concluding part of your research.

Of course, it is important that you show what has been answered by your research, but it is also important to show what has not been answered. If a next researcher would be able to answer this unanswered with new work, it would even make your work stronger and more relevant.

21

Tip 2 – List hypothesis built on your conclusions

Rather than listing unanswered questions underlying your research, you can also list new hypothesis that would be based on your research.

If you have found an interesting conclusion, generating those new hypotheses should be relatively easy. If necessary, you might already relate those new hypotheses to other researchers work. This is likely to make you successful, as other researchers would appreciate the generosity of making their next work easier.

Good and Bad Examples of Cyclical

1 – A Cyclical Example

An example of Cyclical research could basically be any research if it makes sure to add a starting point for new research in the end.

A practical example of this is a 1986 research on treatment of Cancer using chemical antibodies. This research was grounded in another research paper that found out how to create those antibodies. And following those papers, there has been a lot of advances in the field of treating cancer in a very cyclical manner.

Those medical studies attack such large and important topics, that it is impossible to do all the work in one paper. This makes it necessary to do the work step by step and make the scientific knowledge grow with those cyclical, incremental steps.

For other fields of research, the same applies. But sometimes researchers may be tempted to take too much hay on their fork by trying to solve it all at once.

2 – A Non-Cyclical Example

For the example of non-cyclical research, let's take back the example of Artificial Intelligence. The example was a paper that develops a new method for automatically detecting objects on images.

Papers of this type are often very focused on performance: they only try to optimize the number of objects correctly identified. This makes their work difficult to do in a cyclical manner.

Mostly, approaches are relatively similar, and therefore do not make it easy to build on their conclusions. But sometimes there are approaches that are so fundamentally different, that they give rise to a huge number of new researches that try to apply this new approach and benchmark their results.

RULE 7 – BE REPLICABLE

What is replicable research?

Replicable research is research that is very reliable by allowing other research to verify whether re-doing all the data collection would give the same outcome.

Besides other ways for being reliable that have been showed in previous chapters, replicability is the one and only feature that makes your conclusion undeniable.

If it may happen that other researchers do a concurrent research to yours with a different outcome, if you have made sure to incorporate replicability from the start, you are sure to defend your case successfully.

How to make your research replicable?

Tip 1 – Make copying easy

Replicating the complete data collection part of a research can be very time consuming. If your research is easily re-doable, this makes it more likely that your research would be redone.

It is in your favor of allowing easy copying of your research setup, since every time someone would retest your conclusions, this puts your research in the spotlight: a serious contribution to your success.

Tip 2 – Think about replicability before starting

To ensure copy-ability of your data collection, it is important to take this aspect into account from the beginning of your design. It is scientific best practice but also good for the reliability of this specific research.

Topics that take weeks or months of observing phenomena in practice are less likely to reproduce a same output a second time than for example a simple experiment.

Or think about the difference in reproducibility of a survey versus a discussion panel analysis: it would be very hard to find the same discussion panel and make the discussion go the same way, whereas a survey would likely produce more or less the same averages.

Tip 3 – Report execution details

Small details can sometimes have large consequences. You may have heard the anecdote of chaos theory that states that a butterfly clapping its wings in china could cause a tsunami in America?

Small details having an impact on your reproducibility of your research is something that you will want to avoid. The only way to do it is to carefully list all the details of the data collection process.

Simple examples of small details with large impacts are the weather outside that cause different dynamics in a lab experiment. Or image the effect of the moment of collecting surveys in the street: having the survey just after a football match could cause your survey to be filled by tons of football supporters and have a large bias in your conclusion.

Good and Bad Examples of Replicability

1 – A Replicable Example

Research methods that are generally known for their replicability are experimental designs. This feature of experiments is the reason why experiments are very popular.

The reproducibility comes from the fact that the researcher decides which study object would be in which treatment group. This way the researcher can make sure that each group for example has 50% man and women, 50% smokers and non-smokers, etcetera. This will balance out all the unwanted effects.

2 – A Non-Replicable Example

A discussion panel in a case study is an example of something that is much harder to replicate. The researcher has much less control over the data collection in this case, as he cannot decide what people say or not say. There is a lot of randomness in people's way of talking and that may very well influence the discussion.

A way to introduce replicability would be to add more researcher influence on the discussion panel by strictly deciding on the topics that can be discussed, possibly by adding rounds of topics. But at the same time, this may take away freedom to speak for the participants and make the outcomes less interesting.

RULE 8 – BE GENERALIZABLE

What is generalizable research?

Generalizable research is research of which the conclusions are true not only for the case researched, but for any case that may occur.

When researchers work on a paper, it happens that they are so focused on the expected outcomes, that they miss obvious variables that could play a large role in the researched process. This makes the conclusion applicable only in the constrained test environment and this negatively affects the re-usability of the research.

How to make your research generalizable?

Tip 1 – Size your scope and validate

A trap for many researchers in to attack too large research questions. By having a large scope in your research, you often try to answer multiple answers at the same time, and you induce dependencies between those bricks of knowledge.

By looking at smaller bricks of knowledge and posing smaller questions, you can spend more time on testing validity of the generated conclusion under other conditions.

Tip 2 – Test different conditions

The second tip for assuring generalizability is to test different conditions in your research. If you want to find out whether a phenomenon A and a phenomenon B are related, often there may be underlying factors that influence this relationship.

Maybe phenomenon A and B seem positively related in summer, but negatively related in winter. If your research is done in summer, you could be well wrong in winter.

This may seem unrealistic, but such effects unfortunately do occur in practice.

Tip 3 – More evidence is more proof: randomness

A last tip to assure generalizability is to collect multiple data sources for a same research question.

There is a lot of randomness in nature. When you flip a coin 10 times, obtaining more than 8 heads is actually not that rare. And this could strongly bias your conclusion. If you had repeated this experiment three times, it would be much more unlikely to obtain 8 heads each time.

The more evidence you collect, the less likely that your conclusion is biased by random variation.

Good and Bad Examples of Generalizability

1 – A Generalizable Example

As an example of generalizable research, let us look at how to make a survey research more generalizable; Survey research is applied by many graduate students and beginner mistakes are likely.

When doing surveys, it is important to ask many questions about your sample. Students generally have a large network of students and therefore their respondents are often students.

Now this can be a bias: if you collect data on students, your conclusion should be only generalized to the student population. Because there may be a reason that students behave differently than non-students.

To make this generalizable, you could add questions in the questionnaire, to ask whether the respondent is a student or not, and this would allow you to investigate whether students respond differently than non-students.

This way, you can generalize your research to the overall population and make your research more relevant to readers.

2 – A Non-Generalizable Example

As a counterexample of being generalizable, let us have a look at the famous example of the Black Swans. It is an often-cited example and it shows nicely how difficult generalization sometimes can be.

The example dates to a long time ago: to a moment where the world was not as connected as it is now. In those days, the only swans that were known to many where white swans. Simply, black swans occurred only in a part of the world that was not yet discovered.

Imagine the surprise when voyagers started to travel the world and found out that there are also black swans!

This shows that it is possible to know something as soon as we have seen it work, but that the opposite is not true. Seeing a black swan is obviously proof that black swans exist, but not seeing a black swan is no proof that black swans do not exist.

Take this into account into your research and make sure that your conclusions are also valid in other parts of the world. Or else, clearly state the limitations of what you have investigated.

CONCLUSION

Having the 8 rules of research should set you up for success. So before publishing, make sure that your research is attractive, relevant, reliable, incremental, concise, cyclical, replicable and generalizable.

Following each rule in itself is simple. But the hard part is to be it all at the same time.

As examples given throughout the book, it was stated that being too focus on innovation can make you less incremental or that being too focused on reliability can make you less attractive.

The key in all this is balance. Whether your success is defined as getting citations, being on a popular news show, recognition from peers or employers, make sure to balance out each of the 8 rules of this book, and set yourself up for success.

INSPIRATION FOR THIS BOOK

Research methods are an often forgotten, but fundamental part of data science and data analysis. The increasing popularity of data science has inspired Korstanje to share a focus on best practices of research, that can be applied not only in research but in any practice of truth-finding like data science and data analytics. This has been the inspiration the Guide to Successful Research that lies before you.

ABOUT THE AUTHOR

As a dual graduate of Wageningen UR and DSTI, Korstanje has worked on a multitude of projects in research, statistics, data analysis and data science.

To share his experience to aspiring professionals, Korstanje keeps a popular blog on Medium.com, writes books and delivers online courses.

For more resources of this author, check out the following resources:

- linkedin.com/in/jooskorstanje
  For regular updates and content.

- medium.com/@jooskorstanje
  For blogposts on artificial intelligence, data science, research and programming.

- jooskorstanje.com
  For free resources

www.ingramcontent.com/pod-product-compliance
Lightning Source LLC
Chambersburg PA
CBHW030544220526
45463CB00007B/2972